Leeds Lakes
and Ponds

Leeds Lakes and Ponds

Anthony Silson

BEECROFT PUBLICATIONS
2022

Published in 2022 by
Beecroft Publications
72 Waterloo Lane
Bramley, Leeds
LS13 2JF
www.beecroftpublications.co.uk

A CIP catalogue record for this book is available from the
British Library

ISBN 978-0-9930909-8-1
Design and typography by Elizabeth Bee (c) 2022

Printed and bound in the Great Britain by 4edge Limited

To Judith Walker and
Frank and Ivy Mischendahl

CONTENTS

ILLUSTRATIONS

Plates

Figures

Tables

PREFACE

MILL PONDS WERE SUCH A common feature of the 1940s landscapes of Bramley and nearby areas that I, and probably most other residents, came to take them for granted. Similarly today, I suspect most of the inhabitants of Leeds scarcely give much thought to the origins and original functions of the numerous Leeds lakes and ponds that so enhance Leeds' landscapes. This is even evident in the several books written about Leeds where most include only a few lines at best about lakes. My book seeks to rectify this omission.

It aims to show how the spatial distribution of lakes has changed over the last four centuries. How, for example, west Leeds was dotted with an enormous number of ponds during the nineteenth century, whereas today only a handful can be seen. This work also aims to identify the various forms, origins, and functions of Leeds lakes and ponds, not only when they were first made but to discuss any later changes. In so doing, the transient nature of Leeds lakes and ponds should emerge.

In order to achieve these aims detailed case studies of some of Leeds lakes and ponds have been made. The vast majority of Leeds lakes and ponds have been made by people. Hence these lakes are the work's core chapters. But one chapter briefly discusses a few natural or semi-natural lakes.

Because so many of the features have been made by people, both primary and secondary documentary sources, (including internet sources) have been used. Maps have been a major source of information. Alan Godfrey has reprinted the early twentieth century Ordnance Survey maps of most of Leeds. Fortunately those areas which have not been reprinted had, and still have, very few lakes. A great deal of this book was written during the Covid-19 lockdowns, so these maps proved to be invaluable. When restrictions eased then other maps were examined in Leeds Local History Library. On some 1:25,000 Ordnance Survey maps small ponds appear to be shaded completely dark blue whereas most larger ponds are shaded a lighter blue but with a rim of darker blue. There is no difference. According to emails from the National Library of Scotland and the Royal Geographical Society, small ponds are so small that the darker blue rim covers the whole area of the pond.

Fieldwork, over the years, has proved to be a very useful source of information. Much fieldwork was also undertaken in the spring and summer months of 2021. The remaining sources of information have been those given by groups and individuals, either by email or in person.

Unless otherwise stated, when Leeds is discussed its aerial extent is that of the present day Leeds (even though in the past many places lay outside of the then Leeds boundary).

ACKNOWLEDGMENTS

Individuals who have assisted me with information are: Graham Branston, Pippa Hale, Neville Hurworth, Emily Maslen, Hannah Maslen, Isobel Maslen, Matthew Maslen, Steve Murray, Geraldine Usher and Chris Wilson. Pippa Hale has also encouraged me to write this book.

Margaret Plows has accompanied me on many walks in Roundhay Park and thereby re-awakened my interest in lakes.

Particular thanks go to Trevor Plows for his immense computer skills and for changing my rough drawings into the finished products which are visible here.

Institutions which have aided me are: Leeds Local History Library, National Library of Scotland, Royal Geographical Society, and Royal Society for the Protection of Birds.

My thanks also go to Elizabeth Bee for her enormously thorough and skilled editing of this work and for publishing it.

WHEN IS A LAKE A POND AND A POND A LAKE?

WHEN WE WERE VERY YOUNG, many of us enjoyed using twigs and pebbles to try to dam a stream without any worries whether we were making a pond, a pool or a lake. Through experience we began to grasp that these words do not have exactly the same meanings. So I asked the Maslen family, which includes several youngsters of different ages, what the words: ponds, puddles and lakes conveyed to them. Below are their very interesting replies.

Ponds

Have lots of plants in
Like little swimming pools in your garden
You associate them with frogs and lily pads, don't you
You could paddle in one but it would be disgusting
Bigger than puddles
Usually man-made
Walden Pond (Thoreau) (which is actually a kettle lake)

Puddles

Children's play areas
Splashy
Jumping
The image of a child with its arms up, holding a parent's
    hands

Yellow boots/wellies
Children's books
Rainfall
Literally just there from the rain and they dry up very easily
Raincoats
"Sarah and Duck" on CBeebies
Simple
"How to Catch a Star" (book)

Lakes
Swimming
Summer holidays
The islands you get in parks
Where people get their water from
Skimming stones
Camping (like rivers)
Much bigger

But a more formal definition must now be attempted.

Lakes *and* ponds are uncovered, outdoor, enclosed bodies of water found on land. Uncovered reservoirs are included because as landscape features they are no different from lakes or ponds. Even the Ordnance Survey could not make its mind up about whether to designate features as ponds or reservoirs. Up to the 1908 surveys, small bodies of water near mills were called mill ponds. From the surveys of 1915/16 onwards many mill ponds were redesignated as reservoirs. But surprisingly not all. The 1954 survey (1:1,250 map) named the Bramley Wellington Mills' ponds as mill ponds and not reservoirs. As well as lakes, ponds and reservoirs, tarns and moats with water in them are included. However, Leeds has only one tarn, that of Yeadon and locally it is known as Yeadon Dam. And now there are only a couple of moats with water in them throughout the year. By my definition, covered reservoirs and indoor pools of any kind, such as indoor swimming pools, and spilt pools of water are not included. Puddles are not included in this discussion. Except in the driest summers such as that of 1976, lakes, ponds, reservoirs, tarns and moats rarely dry up. In contrast, puddle basins are intermittently wet and dry throughout the year.

There is no clear cut distinction between ponds and lakes. The latter are often thought of as being larger than ponds, but critical values of size have never been agreed. Often lakes are formed by constructing a dam across a valley whereas pond basins are often scooped out of the earth. Whilst this distinction often applies, especially before the twentieth century, there are too many exceptions to make this a workable distinction. Harewood fish *pond* has been formed by damming a valley, whilst Waterloo *Lake*, Roundhay, was described on an 1825 map as a fish pond. As there is no agreed distinction between lakes and ponds I shall use the terms loosely, usually following the designation given on maps. Furthermore, to give some variety I have sometimes used mill pond, and sometimes mill dam. In this publication the words refer to the same features.

NATURAL LAKES AND PONDS

Leeds has few, if any, natural lakes and ponds. There are large semi-circular shaped lakes known as ox bows at the confluence of the Rivers Aire and Calder (Ordnance Survey. 2020). Each lake basin was the former bed of the River Calder so in that sense they were natural. But the cut-offs were made when the Aire and Calder Navigation was constructed or improved.

Plate 1: River Calder ox bows (Google Earth)

Until recently, there was a small natural ox bow lake where Scholebrook Lane crosses the meandering Pudsey Beck at Pudsey. There, two meanders were close together and stream erosion on the outside bend of each meander soon cut through the narrow neck of land between them. The stream then followed a straighter course and its former course was left as an ox bow lake. From memory this lake was in existence by the early 1960s, but during approximately the last twenty years of the twentieth century the shallow water has drained and evaporated away, so only the crumbling banks remain to give a clue that a lake recently existed there. Even on the time scale of human beings, lakes and ponds are so often short-lived features of the landscape.

Yeadon Dam is in part natural, but there is no conclusive evidence that the whole is a natural feature. Its floor is made of impermeable rock which helps retain water (British Geological Survey. 2000). Slightly to the east of the dam there is an outcrop of Rough Rock, which is the geological term for certain layers in the Millstone Grit. Rough Rock is massive but diversified by joints (or cracks) that enable rain water to trickle down to the dam and so help maintain its water level. The latter is also maintained by the dam's mean annual rainfall of about 750mm, which is amongst the highest in Leeds, though no way near the 1,300mm received on the moors to the south west of Keighley (Meteorological Office. 1977).

Despite these advantages, the dam may not be entirely natural. Provided Jefferys' 1775 map is accurate, the dam seems to have been slightly smaller then than today. Yeadon Dam has probably been enlarged. That apart, the crucial question remains: is the barrier holding the water at its southern end natural or has it been made by people? From the geological map the only natural matter that could form a barrier appears to be some lateral moraine, and it is not quite clear whether this occurs at the lake's exit. Alternatively, there could have been a shallow hollow in the till surface, but it is equally not clear that was the case. One of the related problems is that so much activity by human beings has occurred in the southern part of the dam area that obtaining the exact nature of the barrier is

Plate 2: Yeadon Dam

difficult. Yeadon Dam may be a natural feature, but as yet some element of doubt remains.

Middleton Park History Trail (no author, no date) states that Middleton Park Lake is "a naturally occurring feature that has been progressively landscaped over the years". No evidence is given that this was a natural lake, and there appears to be no reason why a lake should form naturally at this spot.

Plate 3: Middleton Park Lake

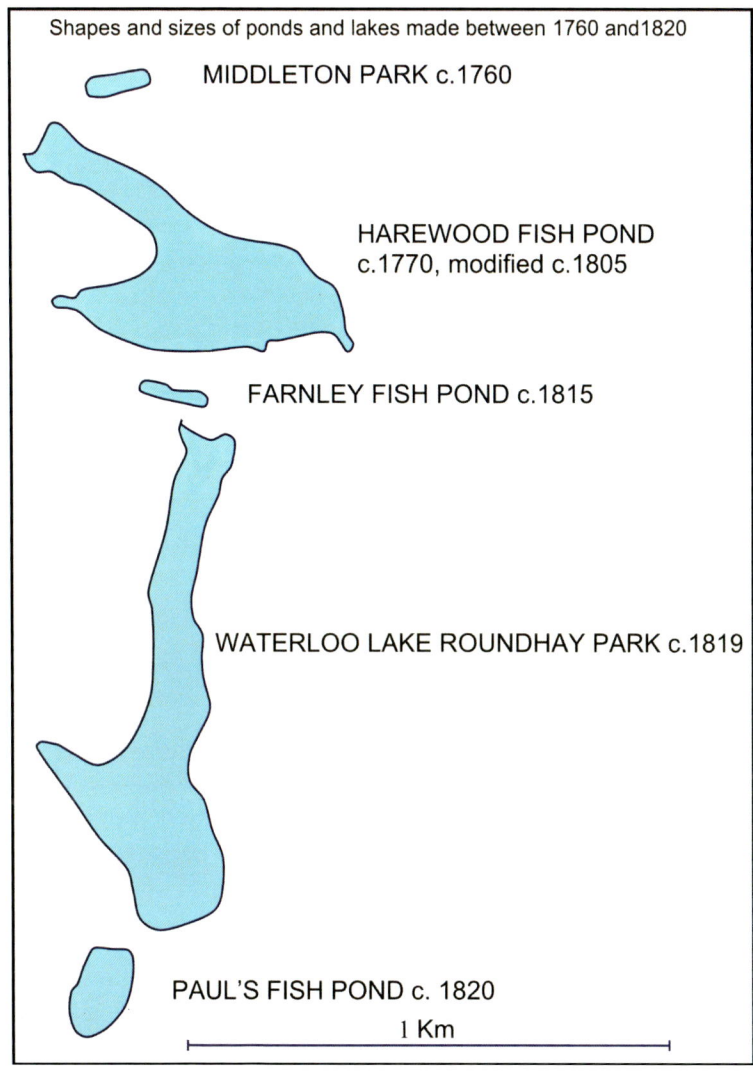

Figure 1: Comparative sizes and shapes of some lakes

– 3 –

PRE-TWENTIETH CENTURY FISH PONDS AND ORNAMENTAL PONDS

JUST TO THE SOUTH OF Harewood House there is a lake or fish pond that was the largest in Leeds (as delimited in 2022) in 1775 (Jefferys. 1775). It had only been recently made, for in 1767 the Reverend Ismay wrote that the valley below the gardens at Harewood House was planned to be a lake (Sheeran. 1980). In 1753, Edwin Lascelles had inherited an estate at Harewood that included Gawthorpe Hall, a nearby small lake and many acres of cultivated enclosed fields (Mauchline. 1993). Edwin was determined to build a new and improved house to replace Gawthorpe, and to change the cultivated land into parkland in which the then existing field boundaries would be eradicated. The new park included a landscape garden with a lake larger than that which already existed. Edwin obtained the services of Lancelot Capability Brown, probably the most well-known landscape designer in England at that time, to plan the changes. Brown was famous for advocating curvaceous landscapes. The new lake at Harewood House therefore had a gentle bow shape in plan view. That is the form depicted not only on Jeffreys' map but on Butterworth's 1797 map of 'near ten miles round Leeds'. By that time Brown's influence was beginning to wane. More picturesque landscape gardens with more trees, more angularity, and more dramatic vistas were being sought. John Loudon was hired to plan changes to the form of the lake. Edwin Lascelles died in 1795 so it was under Edward Lascelles that John Loudon's plans were put into effect very early in the nineteenth century. Loudon's plan was in accord with the changing taste for a more picturesque type of landscape garden. The lake lost its smoothly curvaceous form and instead, in the west, acquired the open crocodile's mouth shape that is still visible today. It was initially called a fish pond and it is still called a fish pond today.

As well as the large pond, small ponds were created by widening, at intervals, a small stream draining into the fish pond from the south. These ponds were designed to give visitors gasps of delight as they explored the pleasure grounds.

Fish ponds have been a feature of the landscape since at least medieval times. Traces of a medieval fish pond are visible on the Cock Beck east of Barwick in Elmet. There is documentary evidence of fish ponds in the twelfth century at Pool, at Calverley in about 1290 and at Rothwell in 1341 (Faull and Moorhouse. 1981). The one at Calverley probably lay at the eastern end of the village near what is now the main road. Historically, Leeds people were a Christian community who did not eat meat on a Friday or during Lent. At those times fish could be eaten and so were an important part of diets, at least for the rich and powerful. The poor probably had a very basic diet based on bread and vegetables with almost no meat or fish (Rothwell and District Historical Society. 2006). Unfortunately, Leeds is mid-way between the Irish Sea and the North Sea, and before the coming of the railways transport was slow, so fresh fish moved long distances risked going bad. As a result, salted fish was sometimes eaten in Leeds rather than the fresh fish which had been landed at seaports. People who wanted fresh fish either had to rely on those caught in rivers and streams, or had to breed their own in fish ponds.

At some point, medieval manorial fish ponds seem to have fallen into disuse, but new fish ponds were made between c.1700 and 1850, for similar reasons to those of medieval times. Like the earlier ponds these new ones were well scattered, but there seem to have been a greater proportion in the east than the west. This is because fish ponds were associated with wealthy people living on large estates and there were more of these with their landscape gardens in the east. Harewood is an exception in that it lies in the north, mid-way between east and west. Fish ponds found in the east included those at Oulton, Roundhay, Swillington House and Kippax. The two latter ponds no longer exist, the one at Kippax lost to coal mining. Oulton Park had ponds before 1800, but the current ponds, including one fish pond, were made in the early nineteenth century.

The fish ponds at Horsforth Hall Park seem to have been the only ones in west Leeds at this period. In 1746-48 there were an astonishing eight fish ponds at Horsforth Hall Park (Read. 2002). Some ponds, especially those in the north, provided fresh water to help prevent other ponds from becoming stagnant. Others, well away from the Hall, were fish breeding ponds. When the fish had matured, the fish from the more distant ponds were moved to ponds near the kitchen. There, some would then be caught to be cooked in the kitchen. These ponds near the house had fountains, partly for ornament but also to help the water from stagnating. Several of the Horsforth Hall ponds had approximately rectangular shapes. Horsforth New Hall, as it was named when it had been built in 1699, replaced a smaller house built in the 1630s on the same site. Whilst the New Hall was being built landscape gardens, which included fish ponds, were created. Their design may have been influenced by similar gardens that already existed at Swillington House, because John Stanhope, the owner of Horsforth New Hall, had married Mary Lowther of Swillington House in 1697. Whether Horsforth Hall grounds were or were not influenced by those at Swillington, such straight-edged lakes were typical of this period of landscape garden design. Gardens of this type are called formal gardens. They preceded the curvaceous style of Capability Brown which itself gave way to a more picturesque style. None of the Horsforth Hall ponds now exist. In 1940, all the remaining ponds in the grounds were filled in so that Nazi bombers would have no navigational aid to bomb the aircraft factory in Yeadon. A small pond in the south-eastern corner of Hunger Hills Woods survived until the 1960s.

Fortunately, Bramham Park remains as a fine example of a formal landscape garden possessing straight intersecting avenues and a straight-edged narrow T-shaped canal. At one time this canal fed water to five ponds, the largest of which is known as the Obelisk Pond (McCracken. n.d.). The garden, like the house, was probably designed by the first owner, Robert Benson, aided by an architect called Thomas Archer. The T-canal was probably made in the 1730s. This was after the ponds had been made at Horsforth Hall Park. However, Benson does not seem to have been influenced by the lakes at Horsforth. Rather, it is believed that he was influenced by similar lakes in seventeenth-century French gardens. Benson had observed these whilst making the grand tour of Europe which so many rich young men then undertook.

The more formal type of lake design may also be seen in the lake in Middleton Park. The lake's small size and its straight-edged rectangular shape were common features of the more formal style of ponds. I have already mentioned that an unknown source claims it is a natural lake. In John de Loghe's will of 1361 a water mill is listed (Newbould. n.d). But there is no mention of a mill pond and no location is given. The elevated site of the present pond, and its form strongly suggest this was not the site of the water mill mentioned in the will. Whether the lake was ever natural is most uncertain, but what is seen today is essentially a human-made eighteenth-century lake. Indeed, there is a map called: *The Township of Middleton in the 1700s* which depicts this lake mainly as it is seen today (Rogers. n.d.). Of course, there remains a slight possibility that the lake is a very modified natural lake. In 1998, the lake was renovated to encourage people to go fishing in inner city areas (*Middleton Park Lake.* 1998). There may have been earlier renovations but if so the essence of the pond has been retained, so an eighteenth-century lake can still be seen. *Middleton Park History Trail* (op. cit.) goes on to state that in the early eighteenth century it was described as a fish pond, the presence of which is again typical of those times.

In the early twentieth century, on winters' evenings when the lake froze, the lake was lit by lanterns and the game of curling was played. Whilst in the summer, rowing boats could be hired for a row across the lake (*Middleton Park History Trail.* op. cit.). Today, it is one of the oldest lakes in Leeds, and it is a little gem to walk round and admire.

The Canal Gardens at Roundhay Park are a much later example of this formal style, being made in the 1830s, by which time the picturesque movement was in full sway (Burt. n.d.). Roundhay Park was created by Thomas Nicholson between 1803 and 1819 (Hurworth. 2005). The park is at least as famous for its Upper Lake and Waterloo Lake as it is for its Canal Gardens. But in addition, by 1811, there were two small fish ponds located between the present Upper Lake and the present Waterloo Lake (Hurworth. 2009). Today, only slight traces of these two fish ponds can be seen in the field. The date when the Upper Lake was constructed is not known, but it is possible that it was made at the same time as the small fish ponds because it would have been a source of fresh water for the ponds. Separating the Upper Lake from the fish ponds is a very impressive cascade that is in

Plate 4: Upper Lake, Roundhay

accord with the picturesque style of the times. The style is also visible in a fake ruined castle.

The valley that was to hold the water in Waterloo Lake was deepened, and the dam to hold the water was built, by soldiers who were unemployed when they returned to England after the Napoleonic Wars (Burt. op. cit.). Nicholson was a Quaker and, in finding work for these men, Nicholson was giving the kind of valuable service that might be expected from a person of that belief. But I suspect too that Nicholson found their labour was comparatively cheap because they would have otherwise been unemployed. Whatever, this merely explains how the work was done and does not explain why Nicholson wanted so large a lake. Edward Armitage, of Farnley Hall, also helped the unemployed by having a fish pond dug on part of his land. But Farnley Fish Pond was tiny in comparison to Nicholson's lake. As indeed was the fish pond, known as Paul's Pond, at Adel, and made by Richard Wormald in 1820 (*Adel Neighbourhood Design Statement*. 2006). This was almost the same time as that of Waterloo Lake which was made between 1815 and 1819. There is another strong reason why it is unlikely that Waterloo Lake was merely created to provide work for the unemployed. This reason goes back to 1803 when Nicholson was buying the western half of the Roundhay estate and a Samuel Elam was buying the eastern half. The boundary between the estates ran down the centre of a valley which Nicholson wished to drown. Until Nicholson bought the eastern valley side, a lake was not possible. With the death of Samuel Elam, his estate passed to Robert Elam. On 25 August 1815, Thomas Nicholson bought the eastern valley side from Robert Elam (W.R.R.D. GD 642 787). Now it is just possible that immediately after the Battle of Waterloo, on 22 June 1815, Nicholson realised that soldiers returning from the battle would be unemployed; that he could quickly persuade Elam to sell, and that all the legal procedures could be accomplished by 25 August; but the time scale is very tight indeed. One major crucial fact is missing. It is not known when Nicholson (possibly with pressure from his wife Elizabeth) decided to create Waterloo Lake. If it was before June 1815 then the presence of unemployed soldiers after this date is irrelevant as to why Nicholson chose to make this large lake. The acquisition of both sides of the valley enabled Nicholson to make a lake in the valley. It is a moderately deep valley which may have made it cheaper to construct a deep lake but its form did not force Nicholson to make so large and so deep a lake.

Weighing up the evidence, it seems more likely than not that Nicholson merely took advantage of a moderately deep valley and of a relatively cheap labour supply and neither was the *reason* for making the lake. If this is indeed the case, the major question then becomes why did Nicholson want such a large lake? A word of caution here. We may never be able to prove his (or his and his wife's) motives, but there are three which seem the most likely. An 1825 map describes Waterloo Lake as a fish pond (Hurworth. 2009. op. cit.). One reason for the existence of the lake is probably the need for fish, especially if Nicholson was a keen fisherman. But think of these facts. There were already two fish ponds in the park and these were nearer the Mansion House (when it was completed and where the fish would be eaten) than Waterloo Lake. Moreover, Waterloo Lake is so large that there would probably have been enough fish to feed the whole of Roundhay at that time.

Plate 5: Waterloo Lake, Roundhay

So even if the lake was partly built for fish there must be other reasons. One of these is that Nicholson liked to have water features near him. According to a Charles Mills, and quoted in Hurworth's 2009 article, Nicholson "used to say that he had made his money out of water and that he was going to have water around him". Another, and equally strong reason, for creating Waterloo Lake could have been to have a bigger lake than that of Lascelles at Harewood. Waterloo is just that bit larger than Harewood's Fish Pond. The areas are Waterloo: 12.98 hectares and Harewood Pond: 10.74 hectares. Two of the richest families in the area would have known each other and their estates. After all, there was a turnpike road between Harewood and Chapel Allerton, and from there it was only a short distance to Roundhay. The families may have known each other before Nicholson acquired his share of the Roundhay estate, as he had bought an estate in Chapel Allerton in 1799 on his return to Yorkshire from London. Assuming Nicholson aimed to have the largest lake in the area, then he had only Harewood Fish Pond to beat. For example, the Avenue Lakes at Temple Newsam, made by William Etty in 1711-12, were mere tiddlers compared to Harewood Fish Pond. And even Gledhow Valley Woods pond was smaller than Harewood. Consider too Edwin Wainhouse's Tower at Halifax (Slack. 1984). It is believed that he erected this tower to get the better of his neighbour, Sir Henry Edwards. If Wainhouse could spend £15,000 just for that purpose, then it is very plausible that Nicholson could also spend £15,000 (Waterloo Lake's cost of construction) at least partly to rival Lascelles.

It is possible that it is mere chance that Waterloo is the larger of the two lakes but even if it is, such a large lake was announcing to Leeds society, and even to Yorkshire society, that the Nicholsons had arrived; they were people of importance. For years, Waterloo Lake had prominence on any map of Leeds, and even today it still stands out on present day Ordnance Survey maps. From 1821 Waterloo Lake would be the largest within Leeds, as delimited by its current boundaries, for over the next fifty years. It was then, and still remains, the largest family-made lake in Leeds.

Roundhay Park estate was up for sale in 1871. It was bought by the Mayor of Leeds, John Barran, with the intention of selling it to Leeds Town Council to use as a large public park (Burt. op. cit.). Barran had to buy the Park himself because the council was not legally allowed to spend the £139,000 for which the estate had been sold until an Act of Parliament was passed to enable it to spend this amount of money. It did so in 1872 and Roundhay Park then became a public park. Later, an outdoor swimming pool was made below Waterloo Lake (Burt. op. cit.). It was popular on hot days, but expensive to maintain and was demolished in probably the 1960s. Rowing boats could be hired for boating on Waterloo Lake. This was another very popular activity, and one our family often engaged in when I was a child. The lakes were possibly then, and still are, the biggest permanent attraction in Roundhay Park. Certainly it was their presence that mostly drew us to visit from far away Bramley when I was young. Today, the margin of Waterloo Lake is somewhere pleasant to take exercise on foot, and the Upper Lake, with its fountain, something to admire. There is also a lakeside cafe.

After Roundhay Park, Leeds never had any new private estates of that great size. But small estates continued to be

Plate 6: Woodhall Lake

developed. One such was that of Daniel Peckover (of a banking family). Peckover built a house called Woodhall Grange on Woodhall Lane, on the borders of Leeds and Bradford (Sheeran. op. cit.). Much of the Grange's garden was occupied by an irregularly roughly rounded-shaped lake that is today located about half a mile west of New Pudsey railway station. The house has been demolished but the lake remains as an oasis of peace between the two busy roads of Woodhall Lane and Bradford Road.

Lakes in Council Parks
In the late nineteenth century, local councils created a very few ornamental lakes in the parks that they themselves had made. East End Park was made by Leeds City Council gardeners (Allsop and Fletcher. c.1908). The southern end of the park initially consisted of large mounds of refuse from pits, but the gardeners were skilled enough to change this unpromising stuff into gardens, and, whilst they were at it, also made a lake. It is possible that it was the initially irregular terrain that prompted the council to make a lake in this park, as lakes and ponds were rather uncommon in Leeds parks made by councils. Of course there were ponds in Roundhay Park inherited from when it was a private estate. Outside the Leeds boundary, as it was in the late nineteenth century, the local council at Pudsey opened a park in 1889 (Pudsey [& District] Civic Society. 2021). The park included a heart-shaped pond near the present bowling green. The pond was small in diameter and was very shallow, even so there were a few accidents. As a result it was decided, in 1935, to drain the pond and to fill in the basin to grow plants.

– 4 –

RESERVOIRS AND MILL PONDS

I F THERE IS ANY DISTINCTION at all between reservoirs and mill ponds it is this. Reservoirs store water for any purpose, including water for drinking and sewage, whereas mill ponds initially stored water for manufacturing. Occasionally mill ponds are, and were, used for other purposes, but not for drinking water. As previously indicated, the Ordnance Survey changed its mind redesignating mill ponds as reservoirs on later editions of some maps. Most of Leeds reservoirs and mill ponds were made during the nineteenth century. Reservoirs were a consequence of the huge growth in population attendant upon an equally large growth in manufacturing in Leeds at this period, whilst mill ponds were one of the agents that permitted such a great industrial growth. Mills and factories needed water to power new machines, such as power looms, and water was needed in processing, for example, in the fulling of cloth. From the mid-nineteenth century, the great domestic demand was increasingly met from mains water, of which more and more had been stored in reservoirs.

Reservoirs
For many years the River Aire was the main source of water in Leeds township (Sellers. 1997). Even the small reservoir made in the late seventeenth century, and located near St. John's Church, had been fed by water taken from the River Aire. However,

in some other parts of Leeds, including Bramley, water was obtained from wells that tapped underground sources (Silson. 2020). By 1830, the River Aire was still a main source of drinking and washing water in those districts near the river. But the River Aire and its tributaries were becoming more and more polluted with mill, factory and human waste. The population had grown so quickly that workers were very often crammed into small houses in narrow yards. There were few or no toilets. Human waste gathered in the yards, though some people were fortunate enough to have the waste periodically removed by scavengers. It is scarcely surprising that under such conditions people were subject to various diseases. Cholera broke out in 1832 and again in 1849. It slowly became accepted that fresh water was needed for drinking and to flush toilets in order to prevent cholera and reduce the risk of other diseases.

A start to obtain fresh water was made in 1834 when H. R. Abraham proposed the construction of a reservoir at Eccup. From there water would reach Leeds by means of a Seven Arch

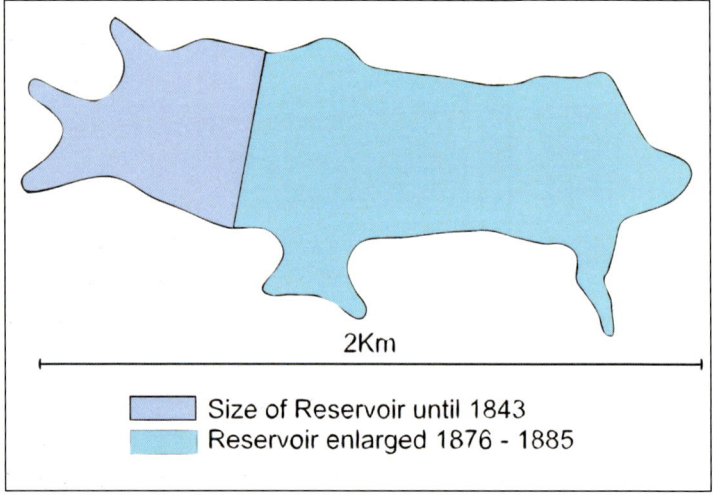

Figure 2: Growth of Eccup Reservoir

Aqueduct to be built at Adel (Broadhead. 1983). An Act of 1837 enabled the Leeds Waterworks Company to put Abraham's proposal into practice. Work began in 1840 and three years later a reservoir had been made at Eccup by damming an existing eastward flowing stream. In 1847, the embankment around the reservoir was raised and the capacity of the reservoir increased. A stream from the north then fed more water into the reservoir. Eccup reservoir then had about a third of its present day area. In 1852, the Leeds Waterworks Company was bought by Leeds Council for £225,730 (*A City's Water Supply*. 1926).

The council was making slow progress laying water pipes. Bramley was one of the last places in the then Leeds Borough to be connected: mains water reached the township by 1860. In the 1860s Leeds was obtaining most of its water from the River Wharfe by means of a pumping station at Arthington. (Sellers, op. cit.). It seems the River Wharfe was less polluted than the River Aire. Some of Leeds water was, though, obtained from Eccup. However, Edward Filliter, the council's engineer, expressed his concern, in 1866, about the reliance on only two sources of water. The Leeds Waterworks Act of 1867 allowed Leeds Town Council to build new reservoirs in the Washburn valley and to enlarge Eccup reservoir (Thornton. 2013). Sellers asserts that it was the growing demand by industry for more water that played "no small role" in leading the council to secure the extra supplies from Eccup and the Washburn valley. However, Edward Filliter's concern about the risk of only two sources of water was surely a contributory factor in encouraging the council to seek alternative water sources.

The council set about enlarging Eccup reservoir by building a new dam further downstream. The work commenced in 1876, not a moment too soon, because in 1875 an Act was passed stating that all newly built houses must have piped water and access to a lavatory. Hence the demand for domestic water supplies increased further. The enlargement of the reservoir was completed in 1885, but it was not fully operational until 1898 because of leaks. The reservoir was then of great size, which it has retained to the present day. The lake was the largest in what is now Leeds, and easily surpassed Nicholson's Waterloo Lake, being about twice its area. It retained its premier position throughout most of the twentieth century. A lake formed by flooding St. Aidan's opencast coal mine in 1988 about equalled it. For a time, water from Eccup was sent via the Seven Arches Aqueduct in the Meanwood valley to a treatment plant at Weetwood and thence to reservoirs at Bramley and Woodhouse. Both of these were initially uncovered and formed lakes or ponds in their own right but were later covered over.

Plate 7: Seven Arches Aqueduct, Adel. It is no longer in use. (Margaret Plows)

Today, Eccup Reservoir's main function is still the provision of water for Leeds residents and industries, but it became a Site of Special Scientific Interest in 1987 (*Adel.* op. cit.). It has become internationally known for attracting wildfowl from as far afield as Siberia and Greenland.

Another, but much smaller, reservoir was built in Leeds at Ardsley in the nineteenth century. At that time Ardsley was part of the West Riding County Council (and not Leeds). The reservoir was built to provide Wakefield with fresh water (Ordnance Survey. 1:1,250. East Ardsley).

Whilst Eccup and Ardsley are the two largest reservoirs in Leeds, there are, and have been, smaller ones. Billing Dam, Rawdon still exists, and it was created to supply water to local residents (Branston. 2010). Earlier this century the reservoir was used by anglers, but on a recent visit, in 2021, I found the reservoir completely locked, and no anglers. From Whitebeck Reservoir near Halton, in east Leeds, water was pumped to another and smaller reservoir east of Halton. From there, water flowed to Temple Newsam and, at a later date, to Halton. Whitebeck Reservoir no longer exists (Dickenson. 1998).

Mainly Mill Ponds

Mill ponds have existed for centuries in Leeds. Kirkstall Abbey monks had a large mill pond where the children's playground is now to be found on Vesper Lane (Ordnance Survey. 1905). Rothwell had a corn mill pond in medieval times, which survived until 1967 when, incredibly, it was destroyed. Yet the pond looks impressive with the church forming a pleasant background on a photograph shown in *The History of Rothwell Castle* (Rothwell & District Historical Society. 2006). The 1905 Ordnance Survey map does not identify the Rothwell pond as a mill pond, but the mill pond at Aberford is so designated. Perhaps flour was no longer being made at Rothwell Mill. The map does, though, indicate that the Rothwell pond was formed by constructing a dam across Carlton Beck, just after it had been joined by a tributary. On Adel Beck, two reservoirs were built in this way to aid manufacturing (Shelton. 2000). The first to be constructed on Adel Beck was Adel Dam which stored water for Scotland Mill, a linen mill further down the valley. In 1820, another reservoir, called Adel New Dam, was constructed upstream of the first dam to ensure the lower reservoir was never without water. In the 1930s, Adel New Dam became part of Thompson's Golden Acre recreational park. Boats were sailed on the lake from 1932

Plate 8: Golden Acre, New Adel Dam

to 1939. The lake was drained during the war. After the war, the park was bought by Leeds City Council, which reinstated the lake but not quite to its previous size. Now the pond is occupied by wildlife including moor hens and ducks and it provides a most pleasing spot for people to sit and relax.

Another type of mill pond was a result of making a much greater change to a stream than by just building a dam across the stream. At Cape Mills, Bramley, some of the water from Farsley/ Bagley Beck was diverted to form a goit or mill race to feed a small triangular pond as shown on Taylor's map of 1811. Water

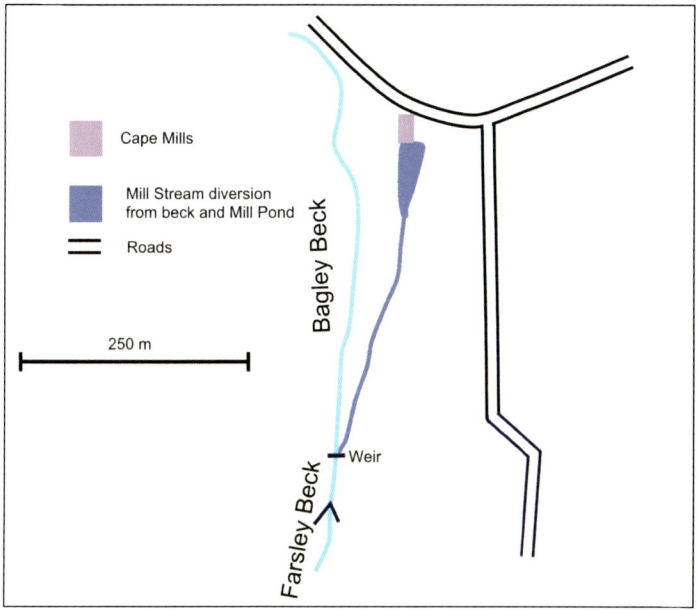

Figure 3: Cape Mills Bramley, goit and pond, 1811

from this pond then drove a water wheel (the site of which was still visible as recently as 1997) to drive the machines in the mill. Cape Mills was built in 1799 as a scribbling and fulling water powered mill. It soon changed to steam power, but the mill pond was still needed to make steam. Whilst the remains of a channel where water was once diverted into the pond may be seen on the Ordnance Survey maps of 1906 and 1915, no water is evident in that channel (Ordnance Survey. Bramley. 1906; Farsley. 1915). In any case the pond probably had insufficient capacity to meet the mill's requirements, for by 1846 another pond had been made near Half Mile Lane and yet another had been added between these two ponds by 1889 (Tithe Map of Bramley. 1846; Ordnance Survey. 1889). These additional ponds were not fed by the goit or stream and as such were of a very common type of mill pond which had a basin that was dug out of the ground. Two other examples of ponds being fed by goits, may still be seen in the Troy area of Horsforth. There, along Oil Mill Beck, two pond basins have been dug out of the valley side. Some water from the beck has been diverted to flow into these basins to form the ponds.

Allen Brigg Mills' (upper) pond, now known as Salter's Pond, is an excellent example of a very common type of mill pond found a great distance from any stream or river. The pond is a rectangle about 600ft (183m) long by 298ft (91m) wide, In April and May 2021, access was denied by two locked gates but at the far south eastern corner there appears to have been a little slipping of the bank whilst the bank at the south-western

Plate 9: Salter's Pond, Pudsey Lowtown

corner has been slightly excavated recently; otherwise the pond appears to be the same shape as on the first edition Ordnance Survey map. A spring feeds the dam with water. The pond is in Pudsey Lowtown but the mill was in Bramley a very short distance away; a township boundary separated the mill and the pond. The straight edges of the pond are typical of many other mill ponds though some are square shaped and a few triangular shaped. One curved edge may sometimes be seen but circular ponds of this type are very rare. The straightness reflects their planned origin on maps. Plate 10 (p. 33), showing Tong Road Mills Dam, very clearly illustrates the nature of a mill pond basin. The size of Salter's Pond is fairly typical too, but there have been a few much larger mill ponds.

Though a few mill ponds were part of tanneries, the majority, like Salter's, were part of textile mills especially those of the woollen, worsted and shoddy trades. Most mill ponds were built at the same time as the mill. Some evidence that this is the case can be found on old maps. For example Belle Isle Mill, Bramley, was built in 1807; Taylor's map of 1811 shows both the mill and a mill pond thus strongly indicating that they were built at the same time (Silson. 2020). There would be no sense in building a pond years before a mill. However, as in the case of Cape Mills, in a few instances extra pond(s) could be added later. Another indication that mill and mill pond were made at the same time can come from documentary sources such as mortgages. John Haley built Waterloo Mills, Bramley, in 1816. When Haley was remortgaging in 1830, a mill pond is included in the items relating to the mill (*Indentures of Mortgage*. 1830). Of course, this is a much greater time interval than the previous Belle Isle example, but it shows the pond was present within fourteen years of the mill being built.

However, there is at least one instance where this principle does not apply. At Springfield Mill in Morley, which opened in about 1864, the two ponds supplying the mill were not built until the end of the nineteenth century (Hindle. 2020). Prior to the ponds being made there was said to be an excellent supply of water that was stored in an enormous cistern in the engine room. The mill's name suggests the water came from a spring, but it is not known why ponds were not used to store the water. Despite this exception, it seems safe to assume the vast majority of ponds were made at the same time as the mill was erected. Provided the mill can be dated then usually the mill pond can also be dated.

A huge number of mill ponds were made during the nineteenth century so that by the early twentieth century Leeds had at least 151 mill ponds (Ordnance Survey. 1905-6, 1915 reprinted by Godfrey). But they were unevenly distributed (Table 1). Most were in an arc from Morley to Pudsey to Calverley-cum-Farsley and then to Guiseley and Yeadon. All this group was just beyond the Leeds boundary at that time. Within

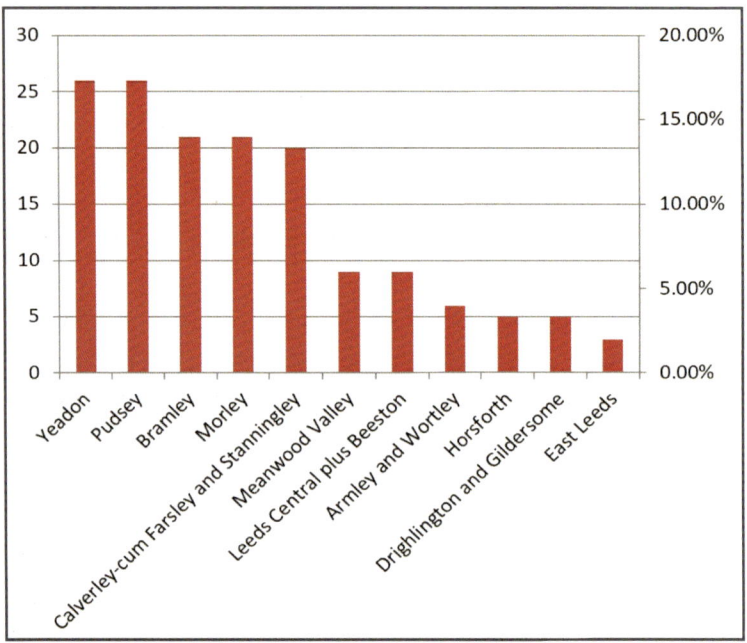

Table 1: Number and percentage share of mill ponds in different parts of Leeds in 1906

Source: Alan Godfrey's Reprints of Ordnance Survey maps of Leeds

the Leeds boundary, only Bramley had large numbers of mill ponds. In contrast east Leeds had only three mill ponds. This concentration in west Leeds occurred because these were the main areas where the woollen (and worsted) textile producing area was located (Silson. 2020).

At the beginning of the nineteenth century east Leeds lacked the growing domestic textile industry found in the west. The central Leeds textile industry was partly being displaced west-ward by increasingly high rents. West Leeds had much land for development and hence lower rents. Its only major handicap was its insufficient water supplies and, as the century progressed, its increasingly polluted water supplies. The becks almost dried up in periods of drought, whilst floods could occur during or just after storms. One such example is Pudsey Beck, known downstream as Farnley Beck. The other problem was a rapid increase in pollution in both the River Aire and its tributaries. Several mill owners commented upon this in a rivers pollution report, 1871 (*The Third Report of the Commissioners. 1871*). Reuben Gaunt of Springfield Mills, Farsley was just one of those who complained about the cost of not being able to obtain all their water from a beck. In Reuben's case it was from Bagley Beck because it was so heavily polluted. Instead, he had to store water that had been pumped up from a well or borehole, in a pond.

Water was crucial both for power (directly or as steam) and for washing wool and finishing the woven cloth. Mill ponds overcame the two major water supply problems by providing an almost constant supply of relatively clean water. Fortunately, these western areas had huge amounts of underground water which could be accessed from wells, boreholes or springs. Wellington Mills, Bramley, derived its water from an artesian source 255ft (77.7m) deep (Yates. 1960). It produced an astonishing

12,000 gallons (54,553l) of water an hour. Another merit of mill ponds was that they could and did provide water at long distances from becks or rivers and so permitted manufacturing development over wide areas. They thus greatly widened the possible sites for mills.

As a consequence of ponds providing reliable sources of relatively clean water, almost all the mills in Pudsey, Calverley-cum-Farsley (including Stanningley) and Bramley, and many elsewhere, that were built *before* the coming of mains water were built with ponds (Table 2). The very few exceptions, such as Union Bridge Mill, Pudsey, were those mills built by a river or a stream. Whether or not Yeadon Dam is a natural feature, it actually served as a mill pond, for in 1871 it was supplying water to Union Mills and Albert Mills in Yeadon (*Third Report*. op. cit.).

Place	Number of mills built before mains water		Number of mills built after mains water	
	With Ponds	Without Ponds	With Ponds	Without Ponds
Bramley	13	3	2	9
Calverley-cum-Farsley and Stanningley	7	0	6	0
Pudsey	10	1	4	10
Morley	8	9	0	6
TOTAL	**38**	**13**	**12**	**25**
Note: A mill could have more than one pond				
Main Sources: Carr 1938; Godfrey reprints of Ordnance Survey Maps; Morley Archives; Pudsey Civic Society 2005; Third Report, 1871; Silson 2020; Strong 2014.				

Table 2: Relationship between mill ponds and water supply

Very few *new* mill ponds were made after the arrival of mains water in an area. There were though a few exceptions. Springfield Mill, Morley has already been mentioned. Two mill ponds were made years after the mill opened. But there were other exceptions. In the cases of Brick Mills and Valley Mills, Pudsey, Holly Bank Mills, Calverley and Springfield Mills, Farsley, the mills may have been planned and/or were under construction just before mains water arrived and so ponds were included in the plans. South Park Mills, Pudsey and Ravenscliffe Mills, Calverley were each in relatively remote locations, probably not connected to mains water for several years after mains water had reached most parts of these settlements. They were both near becks, and it may have been more convenient or even cheaper to store water from the becks in the ponds rather than to try to be connected to mains water.

At Airedale Mills, Rodley, the change to mains water was a long drawn out process. From the company records it is clear that a mill pond occurred during the second half of the nineteenth century and probably from when the mill opened in late 1861 (*Airedale Mills*. 1860-1911). Ten years later, the pond took water from the River Aire despite the river being very polluted. In 1881 it was resolved to use Leeds mains water. However, it is unclear whether this resolution was implemented until much later. But in 1886 borehole water, on the site, was said to be useless on account of the impregnation of strong ochreous pigment. Later that year, samples of water from the River Aire, the dam, and corporation water were taken to compare hardness. Incidentally, the latter was 1.4, the dam water was 3.0 and the river water hardest at 3.4.

In 1889, a new spout was ordered to convey water from the Leeds and Liverpool Canal into the pond. There is confusion here as the canal water can hardly have been clean, and there were complaints in both 1898 and 1903 about customers dirtying the dam water when they were scouring cloth. Yet it was not until 1906 that it again was resolved to use Leeds town water. This time they really must have meant it because in 1907 they were paying £40 per annum for water. This was a huge amount at a time when a foreman was only being paid £63 a year. But it was far cheaper than soap, which cost £212 a year. A few other mills may have dallied in this way.

However, from the *Third Report*, 1871 (op. cit.), it is clear that when mains water became available most mills took up the option to use it with alacrity. Rather surprisingly, most used mains water to supplement the mill's existing water supplies rather than to fully replace it. Waterloo Mills, Pudsey, even continued to obtain water from Tyersal Beck about 98ft (30m) below the mill as well as from wells, boreholes and land drainage. It seems evident from this and other mills that, provided they already possessed mill ponds, these continued to be used in mills for at least some years after mains water was supplied. Waterloo Mills, Bramley, was an exception in that it was using only mains water by 1871. For many years the mill had obtained water from a well in Hough Lane, a quarter of a mile away (Silson. 1991). This unsatisfactory state of affairs was overcome by changing to solely mains water, though the mill pond existed for years. The well house, at the junction of Warrels Road and Hough Lane, was demolished and replaced by a semi-detached house in the 1930s. On at least two occasions the well has opened up again, much to the surprise and consternation of the owner of the house.

Whilst mill ponds were made to assist the manufacturing industry, they have had other uses. Victoria Mills, Bramley, was built by Baptists and from the time the mill opened, in 1837, to the time a new Bramley chapel was built in 1846, people were baptised in Victoria Mills' Dam. Airedale Mills Pond was used by anglers in 1905 and although Bankhouse Mill was demolished about 1890, its large pond survived as a local amenity for almost fifty years (Strong. 2014).

Plate 10: Tong Road Mills Dam, after being drained in 1941
By kind permission of Leeds Libraries. www.leodis.net

During the Second World War, Tong Road Mills Dam, Wortley and Woodhouse Dye Works Pond on Shay Street, became static water tanks (Images. 2021). These were intended to supply water if any incendiary bombs set fire to nearby buildings. New static water tanks, not part of any buildings, were also made in places. There was one down Station Lane (part of Hough Lane), Bramley, and another was made in Woodhouse Square in the centre of Leeds (Silson 2020; Matthews. 2004). They were drained and dismantled soon after the war ended so these ponds had very short lives indeed. So too had a small pond made during the Second World War at Yeadon (Myers. 1989). During the war, Yeadon Dam was drained to hinder enemy bombers' navigation in the area. A huge aircraft factory, said to be the largest under one roof in Europe, was constructed almost where the present airport is now sited. As part of camouflaging the factory a duck pond was made, only to be demolished when the war had been won. Yet there were mill ponds apparently not used that survived for many years. One example is Swinnow Moor Mill Pond which lasted till after the Second World War even though the mill had become an iron foundry in 1911 (Ordnance Survey. 1954; Silson. 2018).

Nevertheless, many mill ponds still appeared to be used by mills even in the post-war period. Wellington Mills, Bramley was still using its pond in 1960. This was because it continued to be a vertical mill in which water was needed to process the cloth, even though the looms were driven by electricity. At Abraham Moon's Netherfield Mills, Guiseley, which is still a vertical mill, water is still supplied from an onsite borehole 800ft (244m) deep (Moon. 2021). However, the mill pond, which presumably was fed from this borehole, was infilled some years ago.

Plate 11: Wellington Mills Pond, Bramley

The condition of the mill boilers also influenced the use of mill ponds. Boilers were often given a great deal of loving care and so they lasted for years. If a boiler was in good working order then it continued to be used and with it the mill pond. It was only in 1948 that Holly Park Dam, Calverley, ceased to be used as a result of the boiler being past repair. Electricity then drove the mill (Pudsey Civic Society. 2005).

From the 1950s onwards, more and more mills closed down and this led to an ever increasing loss of mill ponds so that today very few remain. Mills closed for at least three reasons. Home markets declined as people began to buy less formal wear (Silson. 2020). Imports from third world countries flooded the market as clothes were produced very cheaply in those countries. Exports declined as countries such as New Zealand introduced tariffs.

Crawshaw Mills was the last mill to close in Pudsey. It was demolished and with it went its mill ponds, despite efforts made by Crawshaw School pupils to retain them (Strong. op. cit.). Salter's Pond was earmarked to go the same way, but somehow the campaign to keep this pond was more successful and it survived by the skin of its teeth. Thankfully, it is now, according to Chris Wilson who leases the pond, the subject of a preservation order (Silson. Conversation with Wilson. 2021). The pond is now used for angling. Waterloo Mills, Pudsey, buildings still exist and have a variety of uses. Probably because the mill buildings are still present part of one of the mill ponds is also preserved (Silson. Fieldwork. 2021). Hopefully it still has a long life, though plants are colonising much of the pond. William Gaunt, one of the managing directors of Sunny Bank Mills, Farsley, informed me, on an open day visit, that he is retaining the two remaining ponds at that mill as an amenity.

Rather surprisingly changes in housing have helped to keep some mill ponds. Meanwood Tannery and Winker Green

Mills in Armley have both been converted into flats, and their ponds have been kept to enhance the appearance of the accommodation. Currently, the former Stonebridge Mills, by the Farnley/Wortley Beck, is being converted into flats and houses are being built in its grounds, but the former mill pond will be retained. Low Mills, Rawdon, has been demolished and replaced by houses. The large mill pond has been retained to add some pleasant variety to the housing estate (Silson. Fieldwork. 2021). And the mill pond at Armley Mills has been preserved because the whole mill is now a museum.

Plate 12: Low Mills Pond, Rawdon

– 5 –

THE ROLE OF
EXTRACTIVE INDUSTRIES

QUARRYING AND MINING WERE VERY important Leeds industries, especially from the late eighteenth century to the latter part of the twentieth century. Stone and clay provided materials for the large number of buildings erected during those years. Sand and gravel are used in the construction of roads. For most of that period, coal, which was also extracted from the ground, was a major source of energy. When quarries and surface pits had been worked out, basins were often left which collected rain water and so formed lakes or ponds. Long Close Quarry, near the former Pudsey Lowtown railway station was just one example. As a boy, this former quarry long fascinated me with its craggy sides and its pool of water in the bottom, both of which spelt danger. Today, only a slight contrast in plant cover between the former quarry and its surroundings gives any sign of its former existence.

Contemporary examples of a pond occupying a former quarry occur at Tingley, and Yorkgate Quarry on Otley Chevin.

The floodplain of the River Wharfe at Otley consists of fluvial beds of sand and gravel. These deposits have been worked because they do not include shale deposits, which can sometimes spoil glacial deposits of sand and gravel (Stephens et al. 1953). The built up area of Otley splits the deposits into two, and hence the workings have been in two areas. One is to the west of the bridge over the River Wharfe on the left bank of the river, the other is to the east of the town on the right bank of the river. In addition to the occurrence of the alluvial deposits, two further advantages that encouraged sand and gravel workings were the ease of working on the gently sloping floodplain and the presence of nearby main roads, such as the A659 and the A660, which enabled the sand and gravel to be transported to the many areas where it was to be used.

The eastern workings began in the inter-war period whereas the western workings commenced later, being worked in the 1950s. By that time the eastern workings had almost ended whereas the western workings only stopped in 1996 (Ordnance Survey. Harrogate. 1925; Ordnance Survey. 1958; Stephens. op. cit.). Lakes then formed in the areas that had been worked. A firm called Hanson had worked the western area and this business, along with Leeds City Council, Otley Town Council and Otley Wetlands Nature Reserve Trust, created a nature reserve in the former western workings (Jack. 2008). The nature reserve won a Quarry Producers Association Restoration Award in 2008.

Another example of large lakes stemming from sand and gravel workings is to be seen at Methley Mires, which are part of Leeds Lake District. Sand and gravel were excavated at this site from 1980 until the early 1990s (geograph. n.d.). The basins left by these excavations then became large lakes.

One brickworks where clay was extracted was that of Isaac Chippendale at Scholes (*Scholes Ward.* n.d.). The buildings were destroyed in 1980 but where the clay was extracted is shown today by two small lakes.

Another site where quarrying for clay to make bricks was at, as it eventually became known, Southlands Drive, Chapel Allerton. But this has a more complex history than that at Scholes and whilst the stages in its formation are clear, the motive for the pond's formation is rather uncertain. On a 1767 map of Chapel

Allerton a brick kiln is shown not far from the site of today's pond and the 1767 map also shows a brick pond, though it too is not actually on the site of today's pond (Source of map not given but illustrated in Tucker. 1987). An 1814 map shows a quarry located on or very near the site of the pond today (Taylor. 1814). However, no mention of brick making has been found in early directories between 1821 and 1839 such as those of Baines, Pigot, and Parson and White. The 1851 1:10,560 Ordnance Survey map shows what appears to be a former quarry exactly on the site of the pond. Moreover, this former quarry is located within a couple of fields called brickfields. And the 1889 1:10,560 map also shows a former quarry exactly on the site of today's pond. Though not quite fully proven, there is enough evidence to support the existence of a quarry that was used to extract clay to make bricks.

The site where the clay had been extracted had become infilled by 1906 but is marked by a very small field on the 1906 Ordnance Survey map (Godfrey. reprint. 1906). The land nearby is colonised by rough grass. The site was part of a vast estate called Carr Manor, which in 1925 was owned by Sir Berkeley Moynihan. Digby Chamberlain, in his 1936 obituary of Moynihan, described him as the world's greatest surgeon (Thornton. 2021). Moynihan decided to have the infill excavated to form a basin for a pond. Work to re-excavate the hollow was already underway by 13 January 1926 when a photograph was taken (Annotated photograph. 1926). The original name for the pond was King Lane Pond. That much is clear; the key issue is why Moynihan desired a pond. He probably wanted to create an amenity in the area. Several local press reports, whilst he was alive, state his hobbies were golf and swimming. He was really keen on the latter so that may explain why he wished to have a water feature made as an amenity. But why make an amenity, and why then? By 1925, when the decision to re-excavate the

Plate 13: Southlands Drive Pond, Chapel Allerton

site must have been taken, there were very few houses in the area (Ordnance Survey. Cassini. 1925). For a while the pond probably stood in splendid isolation and even by 1934 very few houses had been built (Ordnance Survey. 1934). One fact is that in August 1926 Moynihan formed a company with himself, a Mr Emsley of Boston Spa and a Mr Mosley of Roundhay as directors to develop part of his estate (*Carr Manor Statement.* 1926). By 1939 many houses had been built in the area near the pond. Moynihan may have predicted that a pond would be an amenity when houses were built. He may have thought such a feature would attract house builders and then house buyers. Another possibility is that Moynihan had the pond made then to be part of Leeds Tercentenary celebrations in 1926. At the moment we do not know and possibly never will know for certain his motives. What is clear is that he made an unique feature in residential suburbia in Leeds and possibly in the whole country in the inter-war period. At one time, Southlands Drive

Pond had world-wide interest on account of a special type of snail that bred there. It remains highly unusual today, but some new housing developments are including former mill ponds such as the one at the former Low Mills. Lakes, which are discussed later, are also part of new housing at Thorpe Park.

Leeds Lake District

Very few lakes were to be seen along the valley of the River Aire between Stourton in the north and Newton in the south in the mid-1960s (*Second Land Utilisation Survey* 675, 676, 687. 1962-68). Yet today there is such a vast number of lakes that this south eastern corner of Leeds merits the designation: Leeds Lake District. This great change in the landscapes of this area has largely arisen through the effects of the extractive industries, especially those of coal mining though, as previously described, sand and gravel extraction at Methley Mires has led to the occurrence of large lakes there.

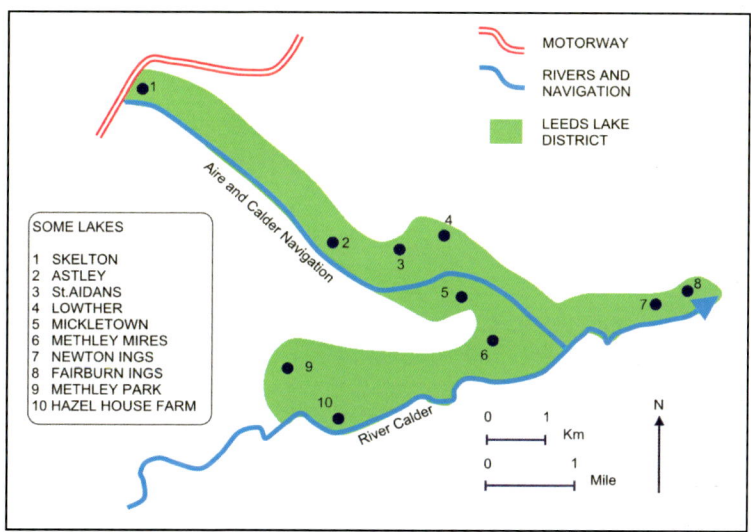

Figure 4: Sketch map of Leeds Lake District

In this south eastern area of Leeds, coal was mined underground or at the surface in opencast mines. The latter were to be found near the banks of the River Aire between Skelton Lake and Allerton Bywater, and a few lakes are a result of former opencast workings. The first stage in opencast coal mining is stripping off the overburden, which includes removing plants and top soil. In the second stage, the coal is loosened by explosives, if it is not already loose, and then dug out. Lastly, the coal is moved from the site possibly initially by conveyors. When the surface or near surface seams are exhausted or for some other reason, such as all the coal that planners have allowed to be extracted has been removed, then mining ceases. A large hole or bowl is left in the ground, which has the potential to become a lake. However, that potential is not always realised. Opencast mining began in 2002 to the west of Clumpcliff Farm at Moss Carr (Silson. Fieldwork. 2003). Mining ended in 2004, and by 2006 the land had been so well restored that it was almost impossible to find any traces except, perhaps, for some young trees that had been planted (Silson. Fieldwork. 2006). The site was good farmland and presumably the farmer wanted the land back to farm. Indeed there is a regulation that restoration should usually occur.

However, in a few cases opencast mining has led to the formation of a lake. A good example is Lowther Lake, Allerton Bywater. This is a large lake secluded by trees and shrubs. Its southern edge is bounded by an oxbow of the River Aire so it is curvaceous, but the northern edge is almost straight. Lowther Colliery, to the north of the present lake, was in existence at the beginning of the twentieth century, and further south a small lake had formed, possibly through subsidence (Ordnance Survey. 1908). By 1958 that lake had either dried up or drained away but the land was marshy so the opencast workings that occurred in the 1970s began on damp land (Brook. 1976; Ordnance Survey.

York, 1974, revised 1958). At the same time that opencast mining was occurring, Allerton Bywater Victoria Anglers were allowed to fish in Willowgarth Lake, just to the west, at St. Aidan's core site (Silson. email. 2021). British Coal wanted to drain Willowgarth to enable them to undertake opencast mining there, so they offered to make Lowther into a lake for the anglers when workings at Lowther ceased. Allerton Bywater Victoria Anglers accepted this offer, and so British Coal made the basin left, when workings had ceased, into a lake. A huge mountain of overburden had accumulated by 1976, some of which may have been deposited in the basin before being flooded. If so, the presence of an island in the lake indicates that the irregular relief occasioned by mining was by no means eradicated. At first,

Plate 14: Lowther Lake

netting was used to catch the fish, but as not too many were caught, Willowgarth was almost drained so all the fish could then be moved into Lowther. Initially, bream were great catches at Lowther but now carp are particularly evident, though bream are still caught. In the case of Lowther, then, opencast mining created a site for a lake but it was the need to enable Allerton Bywater Victoria Anglers to continue fishing that led to the potential for a lake being realised. Lowther Lake is part of Leeds City Council's major nature reserve that stretches from Skelton Lake in the north to Fairburn Ings in the south. Like Astley Lake and the major central area of St. Aidan's, it is managed by the Royal Society for the Protection of Birds (RSPB). Some confusion may exist about the names. Leeds City Council includes part of the former Savile Colliery at Methley, Astley Lake, and Lowther Lake in St. Aidan's (Leeds City Council. 2010). Many people think of St. Aidan's as being only its core or central area.

Plate 15: Opencast mining, St. Aidan's, 2001

St. Aidan's Central Nature Reserve with its large south eastern corner lake and its former opencast workings are well known. As a result, you might think that the large lake is a fine example of opencast mining leading to the formation of a basin and then a lake. Unfortunately you would be wrong. By 1896, there was a small lake in the same position as that of today (Brook. op. cit.). Its origin is not known but subsidence due to mining may have been responsible or it may even have been natural. The lake had become much enlarged by 1932, probably as a result of subsidence stemming from underground coal mining. It cannot have been a result of opencast mining as that did not begin in Leeds until 1942. Though shrunk a little in size by 1950, the lake was essentially still in existence in 1988. For many years since 1950 opencast mining had taken place to the north of the site (Ordnance Survey. Sheet 104: 1958, 1974, 1984). In the 1980s opencast mining moved northwards to such an extent that the small settlement of Astley was devoured. It is now a lost village.

Plate 16: Infilling at St. Aidan's, 2003

In 1988, the River Aire burst its banks and the whole site was flooded to form a huge lake. In a remarkably short time this lake had been drained and opencast mining recommenced in 1997 (Silson. Fieldwork. 2001). It had cost £20 million to re-open, and mining lasted a mere five years and then permanently ceased. Presumably the operation was viable as almost three million tons of coal were excavated. Mining's short life after the floods was a result of the coal seams to the east having been previously worked out, a decision not to mine to greater depths and by

2002 all the coal that could be mined by the current planning permission had been achieved.

When mining finished the gigantic hole that was left was filled in. Lorries carried truck loads of earth from the western area to the eastern part where the deepest basin had been made (Silson. Fieldwork. 2003). There was some sign of less infilling on that part of the site which had been a large lake prior to 1988. In 2001 a plan for St. Aidan's could be seen. This plan had been finalised after years of discussion within a body that included UK Coal (as British Coal was then known) and the local RSPB group. Though there were detailed changes, the plan essentially restored St. Aidan's to what it had been before the 1988 flood, which included the large south eastern lake. The restoration adhered to the plan and so the south east became a lake just as it had been in 1932.

Plate 17: Lake(s) on the site of the former opencast mining, St. Aidan's, 2021

At the northern end of the core or central area of St. Aidan's is a car park and a cafe which has outdoor seating. From some of these seats, the land gently slopes down to the River Aire with large lakes to the left. It is a fine open landscape with extensive views, unlike Lowther which is so sheltered.

On the opposite side of the River Aire, two quite large lakes had formed between 1890 and 1905 to the west of Pit Lane at Mickletown (Brook. op. cit.). These lakes were a result of the land subsiding as a result of coal seams being worked eastwards from Methley Savile Colliery, which had opened in 1873.

By 1932 the northern lake had almost disappeared as colliery waste was dumped into the lake, converting it into a slag heap. Little then changed until 1959 when two narrow lakes had formed to the east of Pit Lane. In 1976, only a sliver of a lake then existed west of Pit Lane as a result of continued tipping, but east of Pit Lane the eastward expansion of underground mining led the new 1959 lakes to expand. So in 1976, to the east of Pit Lane there were four small lakes and one large one. (Brook. op. cit.). By 2000 the large lake had grown slightly at the expense of marsh, but otherwise there had been no changes.

Plate 18: Mickletown lakes

(Ordnance Survey, 2000). That there was so little change was a result of the pit closing in 1985, hence there were no further underground workings.

The 2020 Ordnance Survey map also shows there were almost no changes apart from the further shrinkage of the small lake to the west of Pit Lane, and a new tiny lake that had formed near the Aire and Calder Navigation. The Mickletown lakes are now a Site of Special Scientific Interest because former mining has created saline lakes.

East of Allerton Bywater and on the floodplain on the left bank of the River Aire there are many lakes of which the most well-known is Fairburn Ings. The whole area spatially correlates with where deep coal workings have occurred (Muscroft. n.d.). Like the lakes associated with Savile Colliery, these lakes formed as a result of subsidence following underground coal workings that occurred between 1958 and 1992. This type of mining was occurring in the district during the nineteenth century, and it encouraged marshes to form. In the post-war period, the Coal Board was depositing slag to infill parts of the area. However, a Mr Dickens and a Dr Pickup persuaded the Coal Board and West Riding County Council to have the area designated as a nature reserve. This was achieved in 1957. At that time there were almost no lakes in the area which would become part of Leeds (*Second Land Utilisation Survey*. 676. op. cit.). But as a protected area, subsidence hollows made by deep coal mining from 1958 were no longer infilled and so lakes formed in the hollows. From 1976 the area has been managed by the RSPB. Enormous numbers of birds of different breeds have been spotted in the area as a whole and not just confined to the lakes.

Newton Abbey, in the Fairburn Ings locality, was actually a medieval manor house more correctly known as Newton Wallis (*Fairburn Ings*. n.d.). This house was surrounded by a moat which is still present today and partly still has water in it. This is because it is in part submerged consequent upon subsidence. Apart from this moat, the only moats to partially survive in Leeds are a semi-circular moat half surrounding Guiseley Parish Church, and a couple of small basins, which seem to only contain water after high rainfall, at Scholes. So the Scholes moat is now really a big puddle.

Skelton Lake, at the northern end of Leeds Lake District is, in its history, a lake of two halves, with the northern end lacking a lake until the 1990s. Indeed, the northern part of the landscape, that is now water, scarcely changed from at least 1932 to 1990. Over that period it was farmland crossed by channels, a lane and a pipeline. Part of the southern half already had an old ox bow in the mid-nineteenth century and by 1932 a large lake had formed, which included the former ox bow (Ordnance Survey. 1932). This lake may have formed through underground mining leading to subsidence. It is a possibility but no more than that,

Plate 19: Skelton Lake (Pippa Hale)

though there were three pits in the area: Waterloo Main Colliery, Waterloo Main Colliery (Park Pit), and Rothwell Haigh Colliery (Fanny Pit). The lake was by the River Aire, which is liable to flood, and so any slight subsidence could form a lake.

In the post-war period and up to the mid-seventies, the site was used to deposit sludge from an overhead pipeline (Ordnance Survey. 1956; 1965; 1973; Brook. op. cit.). The source of the sludge is not known, but from the position of the pipeline it seems to have entered from the south, and may, therefore have originated at Stourton Sewage works, which closed after government reorganisation of boundaries in 1974 (Sellers. op. cit.). Wakefield Naturalists complained in 1977 that the site was drying up and asked Yorkshire Water to pump water into the site. However, by 1990, three closely adjacent lakes had formed just south of the River Aire (Ordnance Survey, 1991).

Great changes then occurred within the next ten years. During that decade the M1 extension was constructed along what would become the western margin of the lake basin. It may have been coincidence that this new motorway was constructed at the same time as the present day lake or in some way it may have acted as a catalyst. At any rate, in the late nineties the present day lake came into existence. Between 1990 and 1997, the River Aire had been diverted southwards, the three lakes that existed in 1990 were no more, possibly by having been drained, and the whole site appears as a featureless landscape on the Ordnance Survey map revised in 1997 (and published in 1998). Opencast mining was occurring to the immediate east of the lake site, and it is likely it had spread to what would be the site of the lake. Evidence supporting the former presence of opencast mining on the lake site is: the site was owned by British Coal, the diversion of the River Aire, the eradication of the three existing lakes, a panorama in Rothwell Country Park which states opencast mining occurred on the site of the lake, and a Secret Leeds website which states Skelton Grange was an opencast coal site. The problem with the latter evidence is that no limits of Skelton Grange are given; it may refer to the area immediately east of the present lake. Nevertheless, on balance it does appear likely that the lake occupies an opencast site that briefly existed during the 1990s. During this decade, Colton Beck was diverted to flow into the lake. Initially, it probably formed a source of water to help create the lake. Of the lakes formed recently, Skelton Lake is unusual in having a stream entering it.

Like Astley, Lowther, and St. Aidan's core lakes, Skelton Lake is a nature reserve managed by the RSPB. The lake is probably popular with motorists as it is by the M1 and there is a service station providing refreshment. It is not served by public transport.

There are other lakes, formed after 1945, even in Leeds Lake District, that are not linked to mineral excavations, and many outside its boundaries. Methley Hall is now demolished but part of its park remains. Within the former park a long-established fish pond remains.

To the south east of this pond, on Hazel House Farm, there is a similar shaped pond with straight edges, but rather smaller in size. This small pond is most interesting because it has a very different origin from any lake so far discussed. The Hazel House farmer has deliberately made this pond, along with planting hedges and taking some land out of cultivation, to improve the environment (Silson. Fieldwork. 2013). The pond provides a habitat for wildlife that thrives on or in water such as dragonflies, frogs and newts. The farmer is part of Natural England's stewardship scheme which gives financial compensation for environmental improvement.

Plate 20: Hazel House Farm Pond

Other lakes created in the last sixty years

Lakes and ponds made or used to help wildlife are not restricted to Leeds Lake District. Rodley Nature Reserve occupies land that was part of a sewage works (Silson. 2003). Volunteers leased the site from Yorkshire Water and dug out basins to form lakes and

Plate 21: Rodley Nature Reserve

other wet land. The lake basins were previously farm land which was part of the total area of the sewage works, and is still owned by Yorkshire Water (*Second Land Utilisation Survey.* 687. op. cit.).

An environmental concern of a very different kind has led to a large lake being made some fifty or so years ago in west Leeds, between Butt Lane and Tong Road. It was made by damming Farnley Beck, for this was one of the becks mentioned in the mill ponds section that has huge variations in the amount of water flowing from time to time. Prior to the dam there was sometimes such large quantities of water that flooding occurred. The lake

Plate 22: Farnley Beck Lake

was made to even out the flow of the beck and so reduce the risk of flooding. In February 2022, as a result of melting snow and persistent rain, some flooding occurred downstream of Farnley Beck Lake at Wortley. The A6110 had to be closed after Wortley Beck (also known as Farnley Beck and Pudsey Beck) overflowed its banks (Baron. 2022). But without the retention of water in the lake flooding would have been far more extensive. Farnley

Beck Lake is a very pleasant place to walk round and may act as a nature reserve, but flood control is its main purpose.

The lake nearest Allerton Bywater church was also formed to help control flooding, in this case by the River Aire (Silson email. op. cit.). A new lake is now being planned in the Aire valley, between Apperley Bridge and Kirkstall for the same purpose.

A few purely recreational and ornamental lakes have been made recently. A recreational lake opened about the turn of the century at Tyersal. It was intended to be open to anyone, both as

Plate 23: Tyersal Pond

a picnic site and for anglers. Unfortunately, many people treated the site badly, and so the owner banned general access but still allows anglers to fish there. At None-Go-Byes Farm, about a mile east of the entrance to Leeds Bradford International Airport, ponds have been made for anglers.

A new ornamental pond opened at Roundhay Park earlier this century. It is the main part of the Roundhay Alhambra Gardens and is based on the lake in the original Spanish

Alhambra Gardens. It is a narrow canal feature with very straight edges, and after so many years is a return to the tradition which characterised many lakes and ponds for much of the eighteenth century until Capability Brown made Harewood Fish Pond (Silson. Fieldwork. 2021). It is a very attractive feature, which was intended to be but one of seven copies of major lakes found in different countries. Unfortunately, so far this is the only one to have been created.

The most recent lakes to be made are at Thorpe Park in east Leeds. The opening of the motorway link road in 1999 encouraged building on the land trapped between that motorway, a railway to the north and buildings and roads to the south. Offices and a hotel have been built in the south. This area is followed northwards by The Springs, a retail and entertainment centre. The centre includes an M&S, a restaurant, a takeaway, even a toy shop and a cinema which occupies a most imposing building. In total, there are thus many people employed at Thorpe Park. Further north still, is a residential area still being built. Between this area and The Springs is Central Park, most of which consists of six lakes, though there are also areas of grass, footpaths and seats. No lakes existed in the area prior to 1980 but by 2015 there was a thin lake in the north, a central lake which, from the 2015 Ordnance Survey map appears to be drained, and then in the south a small lake which today is fenced off and so is inaccessible.

Today, the northern lake no longer exists but the central lake site has become one of six lakes now found in the Central Park. So all told, five completely new lakes have been made since 2015. All six lakes are small and to some extent resemble mill ponds, but their shapes are perhaps less regular, though roughly triangular. Bullrushes line the edges of the lakes. The lake basins seem to have been excavated.

Plate 24: One of Thorpe Park's lakes

The lakes visibly demarcate the major change in landscape and land use between The Springs and the residential district. The presence of the ponds may attract house buyers and even increase the value of those residential properties near the lakes. However, the official reasons for the making of the lakes are twofold (Sheridan. 2021). Central Park provides opportunities for Thorpe Park's many workers and visitors and for its residents to take exercise and relax and so promote their physical and mental wellbeing. The other reason for creating the park, with its lakes, is to act as a nature reserve. In so doing, they are thus part of a movement over the last sixty or so years to either create lakes as nature reserves or for lakes formed for other reasons to become nature reserves.

A CLASSIFICATION OF LEEDS LAKES AND PONDS

M Y ATTEMPT AT CLASSIFYING LEEDS lakes and ponds follows on the next page. Examples are given, and you, the reader, may wish to add further examples, provided you do not add these to a library book!

Classification is not without some problems. Perhaps the most serious is classifying lakes at places such as Mickletown Ings and Methley Mires. For example, the owners of Savile Colliery, at Mickletown, had no intention of creating a lake when mining commenced. Lakes formed naturally there through the ground subsiding as coal working advanced. Should the lakes therefore be described as semi-natural or as the nature reserve they became? I have resolved this issue by classifying their initial function as: not planned.

Semi natural:
Rivers Aire and Calder, ox bows, Yeadon Dam.

People made (a selection):

Original Purpose	Method by			
	Dam/Goit	Non-Mineral Excavation	Mineral Excavation	Subsidence
For Fish	Harewood, Waterloo at Roundhay	Tyersal	Lowther	
For General Water Supply	Eccup			
Manufacturing	Adel Dams 1+2, Cape Mills	Salter's Pond		
Defence		Woodhouse Square		
Reduce Risk of Flooding	Farnley Beck			
Ornamental	Waterloo at Roundhay	T canal at Bramham Park, Alhambra Gardens, Southlands Drive (Stage 2)	Southlands Drive (Stage 1)	
Nature Reserve		Rodley, Thorpe Park		
Physical & Mental Well-being		Thorpe Park		
Not Planned			Methley Mires, Otley Bridge End	Mickletown Ings

REFERENCES

n.a. A City's Water Supply. In Hirst S. 1926. *Leeds Tercentenary Handbook*, The Tercentenary Executive, 119-122

n.a. 2006. *Adel Neighbourhood Design Statement*, Leeds City Council

Airedale Mills Company Limited, Business Archive, 1860-1911, University of Leeds Special Collections

Allsop and Fletcher. c.1908. The Public Parks of Leeds. Reprinted in Godfrey, A. *Old Ordnance Survey Maps: Leeds (Osmandthorpe), Sheet 218.07*

Annotated photograph of Southlands Drive Pond, 13 January 1926

Baron, J. 2022. Flood alerts in place for River Aire and Wortley/Pudsey Beck, *West Leeds Dispatch*, 20 February 2022

Branston, G. 2010. *Town Street Rawdon: a village within a suburb*, Branston, G.

Broadhead, I. 1983. *Countryside Walks Around Leeds,* Dalesman Publishing Company

Brook, R. 1976. *The Aire Valley Wetlands,* Brook, R. and Wakefield Naturalist Society

Burt, S. n.d. *An Illustrated History of Roundhay Park,* Burt, S

Carr Manor Estate, Statement by Sir Berkeley Moynihan. *Yorkshire Post,* August 1926

Dickenson, G. 1998. Essay on the back of Godfrey's reprint of *Ordnance Survey Map of East Leeds. 1906, Sheet 218.03*

Fairburn Ings (Newton Abbot) Moat. Historic England listing ID 1009926, English Heritage ID 13285, historicengland.org. uk, accessed 17 February 2021

Faull, M.L. and Moorhouse, S.A. 1981. *West Yorkshire An Archaeological Survey to AD 1500. 1981. Volume 3. The Rural Medieval Landscape,* West Yorkshire Metropolitan Council

geograph: www.geograph.org.uk/photo/1155917, accessed 6 June 2021

Hindle, C. 2020. Hudson Sykes and Bousfield: the rise and fall of a Leeds woollen merchant and manufacturing house. In *Publications of the Thoresby Society,* second series, volume 30

Hurworth, N. 2005. *Thomas and Elizabeth Nicholson The Quaker Founders of Roundhay Park,* Hurworth, N

Hurworth, N. 2009. The Third Lake and other "Fish Ponds" in Roundhay Park. In *Oak Leaves Part Nine,* Oakwood and District Historical Society, 15-17

Indentures of Mortgage from John Haley and others to James Brown. 1830. Privately held

Jack, J. 2008. Former Otley quarry is award-winning nature reserve, *Wharfedale Observer,* 20 July 2008

Leeds City Council. 2010. *Lease of the St Aidan's Trust Land to the Royal Society for the Protection of Birds*

Leeds City Council. 2017. *Skelton Lake, Leeds, LS10*

Leodis images: www.leodis.net, accessed 5 February 2021

McCracken, S. n.d. *Bramham Park, the Historical Tour*

Matthews, F. 2004. *Woodhouse Square, A Leeds Safer Communities Millennium Awards Scheme Project*

Mauchline, H. 1993. *Harewood House,* Moorland Publishing Co. Ltd

n.a. n.d. *Middleton Park History Trail*

Middleton Park Lake. 1998. *Yorkshire Evening Post,* 17 December 1998

Moon, A. n.d. *Scouring, Milling and Finishing.* www.abrahammoon.
co.uk/our-story/the-mill/, accessed 12 May 2021

Morley Mills and the Textile Industry. www.morleyarchives.org.
uk, accessed 20 September 2021

Muscroft, J. n.d. *The Last Colliery in Leeds – Allerton Bywater
1751-1992*

Myers, G. 1989. *Super Factory that built 500 planes in Yorkshire
War,* Yorkshire Post Special Publication

Newbould, J. n.d. *A History of the Township of Middleton in the
Parish of Rothwell Part 1 1066-1750,* manuscript, Leeds Local
History Library

Pudsey Civic Society. 2005. *Calverley in Old Picture Postcards,*
European Library

Pudsey [& District] Civic Society. 2021. Notice board, Pudsey Park

Read, A. 2002. *A History of Hall Park 1700-2000. Horsforth History
Guide No.7,* Horsforth Village Publication

Rogers, I. n.d. *History of Middleton*

Rothwell and District Historical Society. 2006. *Rothwell Castle
and Medieval Life,* RDHS Press

Scholes Ward, www.barwickandscholespc.org/scholes-ward/,
accessed 30 June 2021

Sellers, D. 1997. *Hidden Beneath Our Feet,* Leeds City Council

Sheeran, G. 1980. *Landscape Gardens in West Yorkshire 1680-1880,*
Wakefield Historical Publications

Shelton, T. 2000. *Leeds Golden Acres,* Age Concern Leeds

Sheridan, D. 2021. Architects unveil new park, *Yorkshire Evening
Post,* 29 June 2021

Silson, A. 1991. *Bramley, Half a Century of Change,* Silson
2001. Fieldwork
2002. Fieldwork
2003. Fieldwork
2003. *The Making of the West Yorkshire Landscape,* Wharncliffe

Books
2004. Fieldwork
2006. Fieldwork
2013. Fieldwork
2018. *Some Stanningley Industries*, alsilson.wordpress.com
2020. *Bramley in West Yorkshire 1775 to 2020*, Bramley History Society
2021. Conversation with C. Wilson, tenant of Salter's Millpond
2021. Email correspondence with Councillor S. Murray
2021. Fieldwork
Slack, M. 1984. *Portrait of West Yorkshire*, Robert Hale
Stephens, J. et. al. 1953. *Geology of the Country between Bradford and Skipton*, HMSO
Strong, R. 2014. *Pudsey's Mills A Lost Textile Heritage*, Pudsey and District Civic Society
n.a. 1871. *Third Report of the Commissioners: Appointed in 1868 to inquire into the best means of preventing the pollution of rivers; pollution arising from the woollen manufacture, and processes connected therewith. Volume Two*, HMSO
Thornton, D. 2013. *Leeds, A Historical Dictionary of People Places and Events*, Northern Heritage Publications
Thornton, D. 2021. *Leeds, A Biographical Dictionary*, Beecroft Publications
Tucker J. 1987. *Chapel Allerton Historical and Architectural Trail*, Manpower Services Commission
West Riding Register of Deeds (W.R.R.D.). *GD 642 787 (Elam to Nicholson)*, Wakefield
Yates, W. Ltd. 1960. *Rand Exhibition Easter Show*, privately held

Maps
British Geological Survey. 2000. *Sheet 69, Bradford, Solid and Drift*, 1:50,000

Butterworth. 1797. *A map of near 10 miles around Leeds*
Jefferys T. 1775. *The County of York Survey'd.* Reprinted by Scolar
   Press, Menston, 1971
Meteorological Office. 1977. *Rainfall Averages 1941-70 – Map of*
   Southern Britain
Plan: Township of Bramley. 1846
Second Land Utilisation Survey. 1962- 68. 1:25,000. Sheets 675, 676,
   687. (Survey conducted by A. Coleman, published by The Isle
   of Thanet Geographical Association)
Taylor J. 1811. *Plan of the Township of Bramley*
Taylor J. 1814. *Plan of the Manor and Township of Chapel Allerton*

Ordnance Survey Maps
1:1,250:
1954. *Sheet SE2435 NE, Bramley*

1:2,500:
1905. *Sheet 202.16, Kirkstall.* Godfrey A., reprint
1906. *Sheet 202.15, Bramley.* Godfrey A., reprint
1906. *Sheet 203.10, Chapel Allerton.* Godfrey A., reprint
1915. *Sheet 202.14, Farsley.* Godfrey A., reprint
1934. *Sheet 203.10, Chapel Allerton*

1:10,000:
Sheets SE33 SE Temple Newsam, SE33 SW Stourton. All for the years
   when published: 1956, 1965, 1973, 1991

1:10,560:
1851. *Sheet 203, Chapel Allerton*
1889. *Sheet CCII, SE, Bramley*
1889. *Sheet CCIII, SW, Chapel Allerton*
1889. *Sheet CCXXXIII, SW, East Ardsley*

1908. *Sheet CCXXXIV. NW, Methley*
1932. *Sheet CCXVII, SE, Stourton and Temple Newsam*

1: 25,000:
1974. *Sheet SE23/33, Leeds*
1990. *Sheet 683, Leeds*
2000. *Explorer 289, Leeds*
2015. *Explorer 289, Leeds*
2020. *Explorer 289, Leeds*

1:50,000:
1974 (published date - revision of 1958). *Sheet 105, York*
1974. *Sheet 104, Leeds and Bradford*
1984. *Sheet 104, Leeds and Bradford*
1998. *Sheet 104, Leeds and Bradford*

1:63,360:
1925. *Sheet 26, Harrogate*
1925. *Original sheets 26 and 31.* Reprinted as 104 at a scale of
    1:50,000 by Cassini, as *Leeds and Bradford*
1958. *Sheet 96, Leeds and Bradford*

All 60 of the Ordnance Survey Maps covering the administrative
    boundary of Leeds as it is today and reprinted by Alan Godfrey

INDEX

ABOUT THE AUTHOR

FOR MOST OF HIS EIGHTY plus years, Anthony Silson has been a Leeds resident. At West Leeds High School for Boys he had to choose either geography or history as one of his GCE O level subjects despite enjoying both. He chose geography and this eventually led him to obtain two degrees in geography from the University of Liverpool. Though fully aware that some ponds were made by people, at that period in his life he was especially interested in how lakes were made by natural processes.

After he had obtained a postgraduate certificate in education he taught teens geography in Barnsley, Bradford and Leeds.

Near his retirement, his interest in history was rekindled. He has written and published over thirty articles, including *The Number of Days with Thunder Heard in Bramley, West Yorkshire 1962-2010* (2011), *Barnsley College of Technology and Bradford Technical College in the 1960s* and *Forced Rhubarb in West Yorkshire c.1852-2017* (2019). His published books include: *The Making of the West Yorkshire Landscape* (2003), *The Green Spaces of Bramley* (2006), and *Bramley in West Yorkshire 1775 to 2020* (2020).

It was whilst writing *The Making of the West Yorkshire Landscape* that he realised just how many lakes there were in Leeds and how so many had been recently made.